HOUGHTON MIFFLIN H

MATH
Expressions
Common Core

Dr. Karen C. Fuson

GRADE

K

Volume 2

This material is based upon work supported by the
National Science Foundation
under Grant Numbers
ESI-9816320, REC-9806020, and RED-935373.

Any opinions, findings, and conclusions, or recommendations expressed in this material
are those of the author and do not necessarily reflect the views of the National Science Foundation.

 HOUGHTON MIFFLIN HARCOURT

Homewørk

Name _____

Fill in the partners to complete the partner equation.

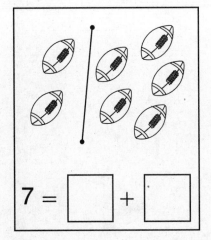

$7 = \boxed{} + \boxed{}$

$7 = \boxed{} + \boxed{}$

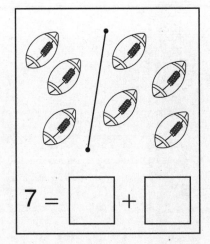

$7 = \boxed{} + \boxed{}$

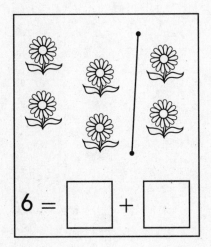

$6 = \boxed{} + \boxed{}$

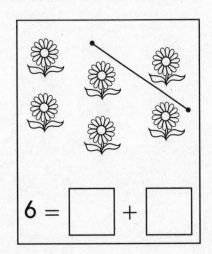

$6 = \boxed{} + \boxed{}$

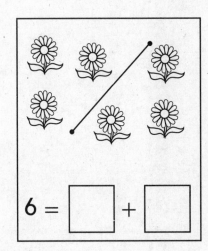

$6 = \boxed{} + \boxed{}$

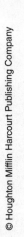

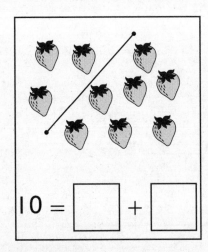

$10 = \boxed{} + \boxed{}$

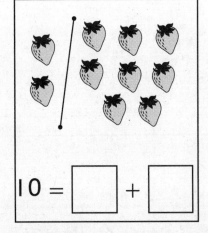

$10 = \boxed{} + \boxed{}$

On the Back Draw a picture for the equation $7 = 6 + 1$.

Find Partners of 10

Homework

Draw a line to show the partners. Write the partners.

10 = [   ] + [   ]

10 = [   ] + [   ]

10 = [   ] + [   ]

10 = [   ] + [   ]

10 = [   ] + [   ]

10 = [   ] + [   ]

10 = [   ] + [   ]

10 = [   ] + [   ]

10 = [   ] + [   ]

Teen Numbers and Equations **69**

Name _____

Draw to show 10 + 2.

Draw to show 10 + 5.

Teen Numbers and Equations

Homework

Write the partners.

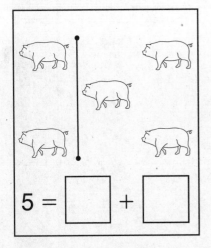

5 = ☐ + ☐

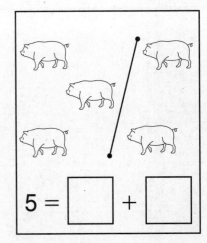

5 = ☐ + ☐

5 = ☐ + ☐

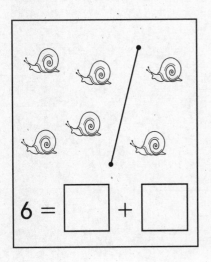

6 = ☐ + ☐

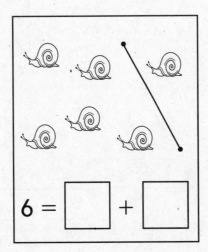

6 = ☐ + ☐

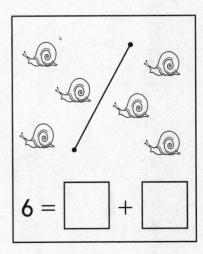

6 = ☐ + ☐

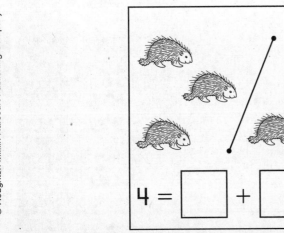

4 = ☐ + ☐

4 = ☐ + ☐

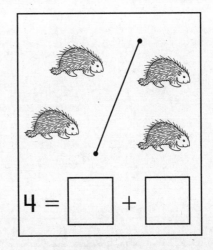

On the Back Draw a picture for the equation 6 = 2 + 4. Draw the Break-Apart Stick.

Addition and Subtraction Stories: Grocery Store Scenario **71**

Addition and Subtraction Stories: Grocery Store Scenario

Name _____

Homework

Write the partners.

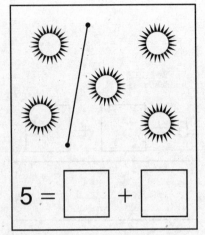

5 = ☐ + ☐

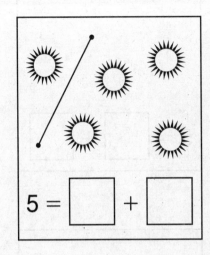

5 = ☐ + ☐

5 = ☐ + ☐

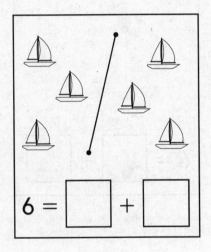

6 = ☐ + ☐

6 = ☐ + ☐

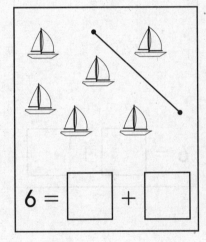

6 = ☐ + ☐

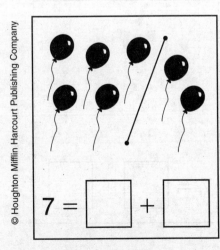

7 = ☐ + ☐

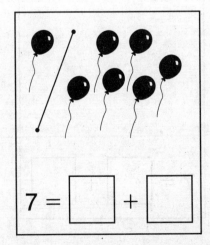

7 = ☐ + ☐

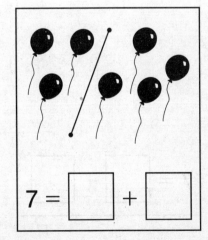

7 = ☐ + ☐

On the Back Make your own partner drawings. Write the partners.

Practice Teen Numbers and Equations **73**

$5 =$ [ ] $+$ [ ]

$5 =$ [ ] $+$ [ ]

$5 =$ [ ] $+$ [ ]

$6 =$ [ ] $+$ [ ]

$6 =$ [ ] $+$ [ ]

$6 =$ [ ] $+$ [ ]

$7 =$ [ ] $+$ [ ]

$7 =$ [ ] $+$ [ ]

$7 =$ [ ] $+$ [ ]

Practice Teen Numbers and Equations

Homework

Draw a line to show two partners. Write the partners.

6 = ☐ + ☐

6 = ☐ + ☐

6 = ☐ + ☐

5 = ☐ + ☐

5 = ☐ + ☐

10 = ☐ + ☐

10 = ☐ + ☐

10 = ☐ + ☐

10 = ☐ + ☐

10 = ☐ + ☐

4 = ☐ + ☐   4 = ☐ + ☐

🔵 **On the Back** Make your own partner drawings. Write the partners.

5 = ☐ + ☐

5 = ☐ + ☐

5 = ☐ + ☐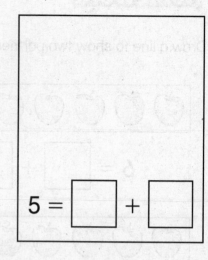

6 = ☐ + ☐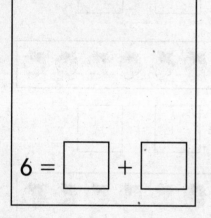

6 = ☐ + ☐

6 = ☐ + ☐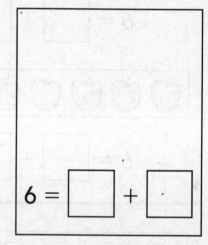

7 = ☐ + ☐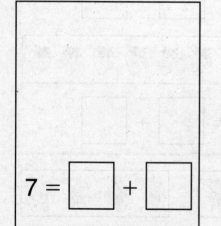

7 = ☐ + ☐

7 = ☐ + ☐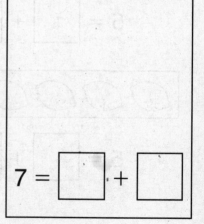

Break-Apart Numbers for 10

Homework

Name _____

Draw Tiny Tumblers on the Math Mountains.

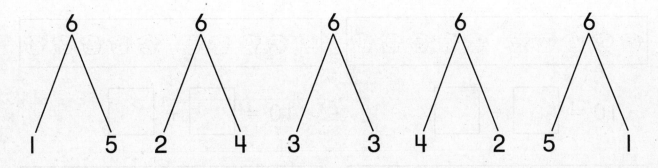

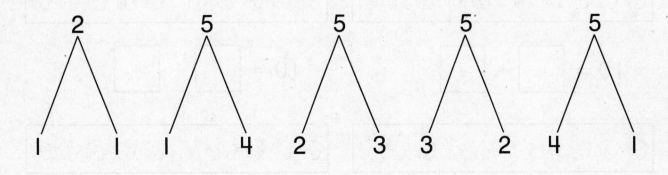

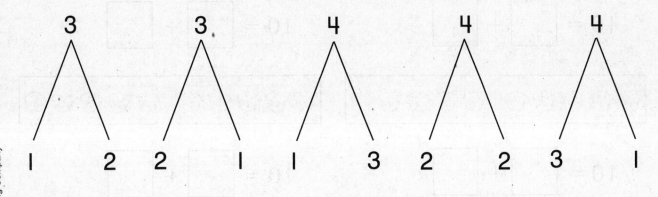

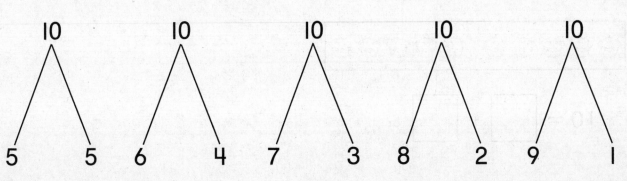

Partners of 10 with 5-Groups **79**

Homework

Draw a line to show the partners. Write the partners.

10 = ☐ + ☐

10 = ☐ + ☐

10 = ☐ + ☐

10 = ☐ + ☐

10 = ☐ + ☐

10 = ☐ + ☐

10 = ☐ + ☐

10 = ☐ + ☐

10 = ☐ + ☐

Partners of 10 with 5-Groups

Homework

Name

Draw Tiny Tumblers on the Math Mountains.

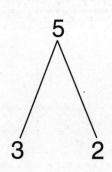
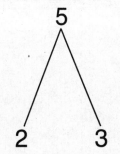
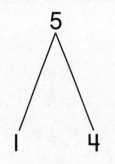

5	5	5	5	2
4 1	3 2	2 3	1 4	1 1

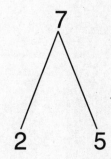

7	7	7	7	7
6 1	5 2	4 3	3 4	2 5

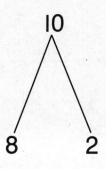
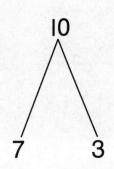
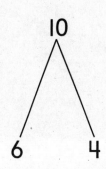
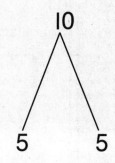

10	10	10	10	10
9 1	8 2	7 3	6 4	5 5

Write the numbers 1 through 20.

On the Back Make a picture with squares, circles, triangles, and rectangles.

Homework

Homework

1. Draw lines to match.

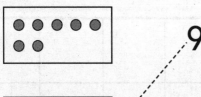   9

   7

   10

   6

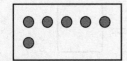   8

2. Make two matches.

   4

   5

   3

   2

   1

3. Connect the dots in order.

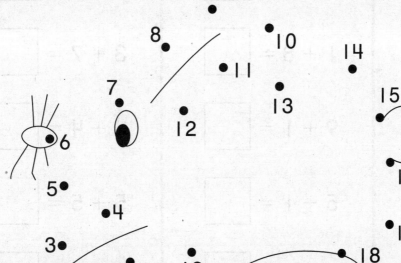

4. Write the numbers 1 through 20.

5. Add the numbers.

$4 + 1 =$ ☐ $2 + 3 =$ ☐ $3 + 1 =$ ☐

$1 + 3 =$ ☐ $3 + 0 =$ ☐ $2 + 1 =$ ☐

$1 + 2 =$ ☐ $2 + 2 =$ ☐ $1 + 3 =$ ☐

$2 + 3 =$ ☐ $3 + 2 =$ ☐ $5 + 0 =$ ☐

$1 + 5 =$ ☐ $1 + 8 =$ ☐ $3 + 7 =$ ☐

$3 + 4 =$ ☐ $9 + 1 =$ ☐ $4 + 4 =$ ☐

$4 + 3 =$ ☐ $5 + 1 =$ ☐ $5 + 5 =$ ☐

$2 + 7 =$ ☐ $6 + 3 =$ ☐ $3 + 5 =$ ☐

Homework

1. Draw Tiny Tumblers on the Math Mountains.

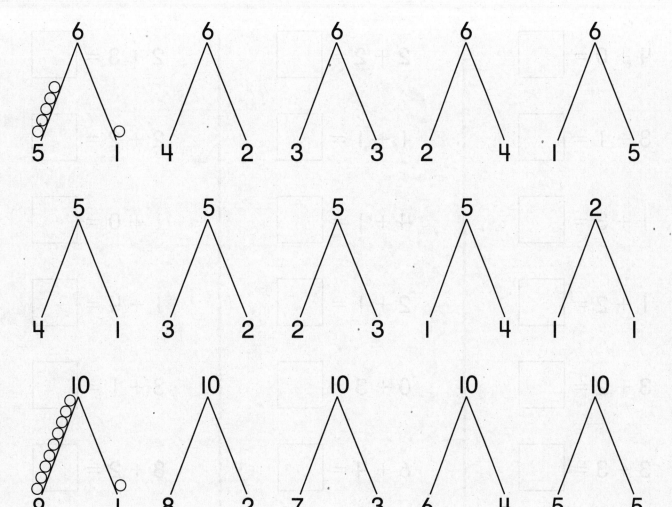

2. Write the numbers 1 through 20.

3. Add the numbers.

$4 + 0 = \square$	$2 + 2 = \square$	$2 + 3 = \square$
$3 + 1 = \square$	$1 + 1 = \square$	$2 + 2 = \square$
$1 + 3 = \square$	$4 + 1 = \square$	$1 + 0 = \square$
$1 + 2 = \square$	$2 + 1 = \square$	$1 + 4 = \square$
$3 + 2 = \square$	$0 + 3 = \square$	$3 + 1 = \square$
$3 + 3 = \square$	$6 + 4 = \square$	$8 + 2 = \square$
$2 + 4 = \square$	$3 + 4 = \square$	$6 + 1 = \square$
$3 + 5 = \square$	$4 + 2 = \square$	$5 + 2 = \square$
$1 + 5 = \square$	$6 + 3 = \square$	$3 + 7 = \square$

Addition Equations

Homework

1. Add the numbers.

2 + 2 = ☐   0 + 2 = ☐   1 + 2 = ☐

3 + 1 = ☐   1 + 4 = ☐   5 + 0 = ☐

2 + 3 = ☐   4 + 0 = ☐   1 + 3 = ☐

2 + 1 = ☐   0 + 5 = ☐   1 + 1 = ☐

3 + 2 = ☐   4 + 1 = ☐   2 + 3 = ☐

2. Connect the dots in order.

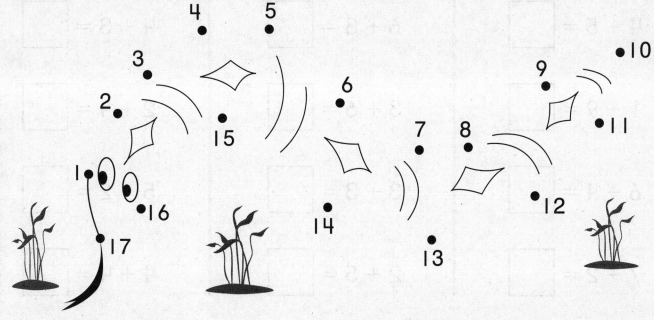

3. Add the numbers.

2 + 2 = ☐   1 + 3 = ☐   5 + 0 = ☐

3 + 0 = ☐   2 + 1 = ☐   4 + 1 = ☐

1 + 4 = ☐   1 + 0 = ☐   1 + 2 = ☐

2 + 3 = ☐   3 + 2 = ☐   4 + 1 = ☐

0 + 4 = ☐   3 + 1 = ☐   1 + 1 = ☐

4 + 5 = ☐   6 + 3 = ☐   4 + 3 = ☐

1 + 9 = ☐   3 + 6 = ☐   2 + 4 = ☐

6 + 4 = ☐   3 + 3 = ☐   5 + 2 = ☐

7 + 2 = ☐   2 + 5 = ☐   4 + 4 = ☐

Write Addition Equations

Homework

Name _____

Count the stars. Write the number.

_____

_____

_____

_____

Homework

Count the gray stars. Write the number.
Count the white stars. Write the number.
Write how many stars in all.

☆ ☆ ☆ ☆ ☆   ☆ ☆ ☆ ☆
☆

_____ + _____ = _____

☆ ☆ ☆ ☆ ☆   ☆ ☆ ☆ ☆ ☆

_____ + _____ = _____

☆ ☆ ☆ ☆ ☆   ☆ ☆
☆ ☆ ☆

_____ + _____ = _____

☆ ☆ ☆ ☆ ☆   ☆ ☆ ☆
☆ ☆

_____ + _____ = _____

Partners of 10: Stars in the Night Sky

Homework

Name _____

Circle the 10-group in the picture.
Write the equation with a group of tens and more ones.

$\underline{\quad 10 \quad}$ + $\underline{\quad 9 \quad}$ = $\underline{\quad 19 \quad}$

_____ + _____ = _____

_____ + _____ = _____

_____ + _____ = _____

_____ + _____ = _____

_____ + _____ = _____

_____ + _____ = _____

_____ + _____ = _____

_____ + _____ = _____

🔵 **On the Back** Choose a teen number. Draw that number of circles. Make a 10-group.

© Houghton Mifflin Harcourt Publishing Company

Solve and Retell Story Problems

Name _____

Practice

Add the numbers. Use your fingers or draw.

4 + 1 = ☐   3 + 3 = ☐   5 + 4 = ☐

3 + 2 = ☐   5 + 3 = ☐   2 + 2 = ☐

2 + 0 = ☐   4 + 2 = ☐   4 + 3 = ☐

5 + 2 = ☐   3 + 1 = ☐   5 + 1 = ☐

2 + 1 = ☐   4 + 0 = ☐   4 + 2 = ☐

Equal or unequal? = or ≠	Equal or unequal? = or ≠	Equal or unequal? = or ≠
3 2 + 1	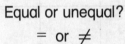	1 0 + 2
2 2 + 2		3 1 + 2
1 3 + 2		2 2 + 3
4 1 + 3		5 4 + 1
5 3 + 0		4 1 + 2

🡒 **On the Back** Write the numbers 1–100.

1	2								10
11									
									100

Make Quantities 1–20

Name _____

Homework

Circle the 10-group. Write the ten and ones in each equation.

___10___ + ___3___ = ___13___   _____ + _____ = _____

_____ + _____ = _____   _____ + _____ = _____

_____ + _____ = _____   _____ + _____ = _____

_____ + _____ = _____   _____ + _____ = _____

_____ + _____ = _____   _____ + _____ = _____

➡ **On the Back** Make a picture with shapes.

Name _____

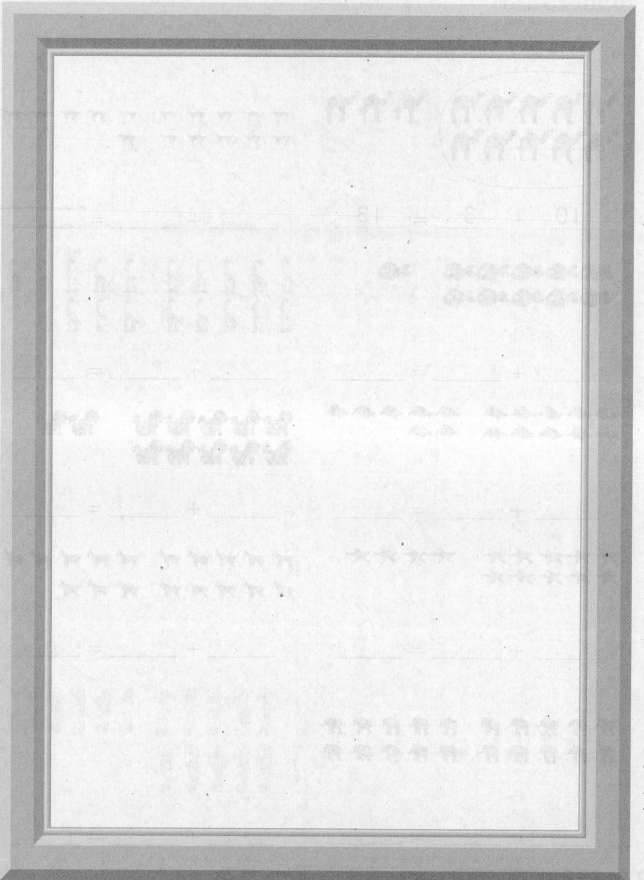

Numbers 1–20

Homework

Draw Tiny Tumblers on the Math Mountains.

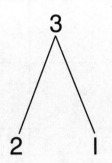

3
1   2

3
2   1

4
3   1

4
2   2

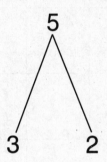
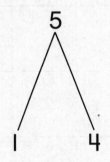
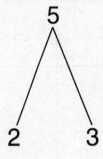

5
4   1

5
3   2

5
1   4

5
2   3

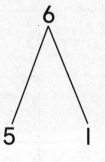
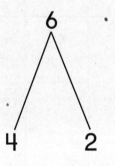
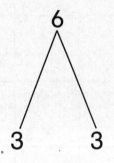
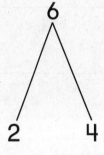

6
5   1

6
4   2

6
3   3

6
2   4

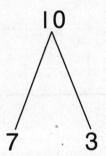

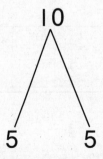

10
9   1

10
7   3

10
6   4

10
5   5

Write the numbers 1–100.

1									
11									
									100

Review Partners

Homework

Draw circles to show each number.
Write the ten and the ones below the circles.

11	12	13	14	15	16	17	18	19 20
10+1	10+	+	+	+	+	+	+	+

Complete the equations.

$$12 = 10 + \underline{\hspace{1cm}}$$
$$14 = 10 + \underline{\hspace{1cm}}$$
$$19 = 10 + \underline{\hspace{1cm}}$$
$$15 = 10 + \underline{\hspace{1cm}}$$

$$16 = 10 + \underline{\hspace{1cm}}$$
$$17 = 10 + \underline{\hspace{1cm}}$$
$$18 = 10 + \underline{\hspace{1cm}}$$
$$13 = 10 + \underline{\hspace{1cm}}$$

On the Back Add the numbers.

Partners of 6, 7, 8, and 9 **99**

1 + 0 = ☐   2 + 1 = ☐   2 + 3 = ☐

3 + 1 = ☐   2 + 2 = ☐   1 + 1 = ☐

4 + 1 = ☐   1 + 2 = ☐   3 + 2 = ☐

3 + 1 = ☐   1 + 4 = ☐   4 + 1 = ☐

1 + 3 = ☐   4 + 0 = ☐   1 + 2 = ☐

4 + 5 = ☐   3 + 4 = ☐   2 + 8 = ☐

1 + 7 = ☐   3 + 7 = ☐   4 + 4 = ☐

9 + 1 = ☐   8 + 2 = ☐   2 + 7 = ☐

1 + 5 = ☐   5 + 4 = ☐   4 + 2 = ☐

Homework

Draw Tiny Tumblers on the Math Mountains.
Write the partners.

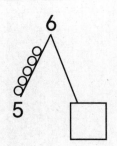 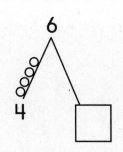 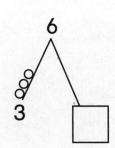 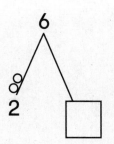 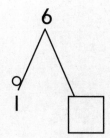

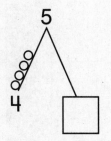 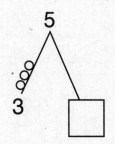 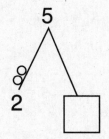 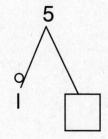 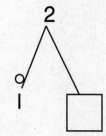

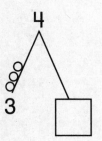 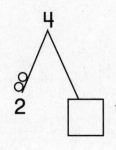 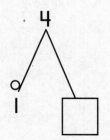 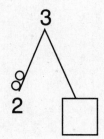 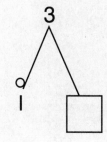

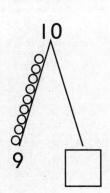 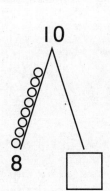 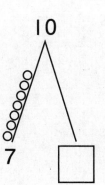 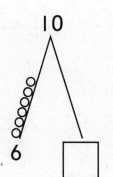 

On the Back Draw four different Math Mountains for 9.

Tens in Teen Numbers: A Game

Homework

Name

Think 5-groups to find the totals.

Color each balloon.

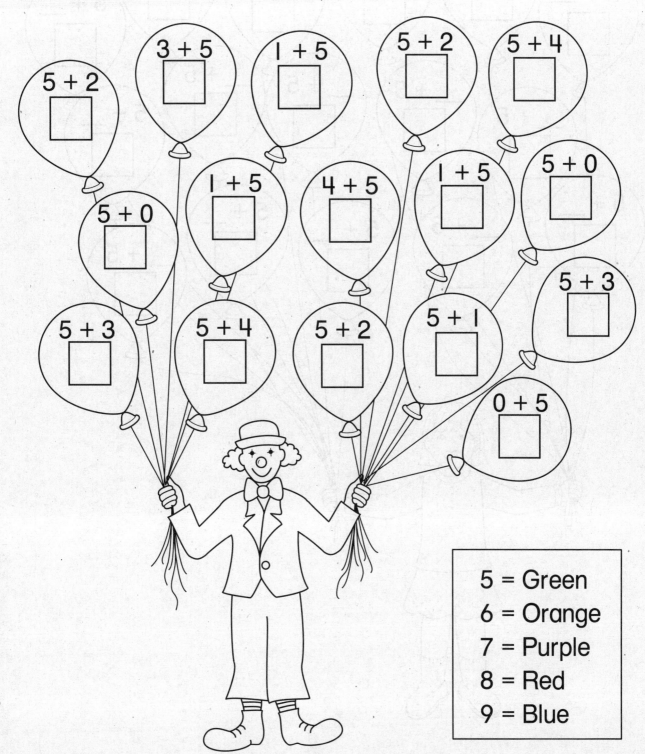

5 = Green
6 = Orange
7 = Purple
8 = Red
9 = Blue

⬤ **On the Back** Make and answer your own 5-group problems.

Practice

Name _____

Subtract the numbers. Use your fingers or draw.

5 – 1 = ☐   5 – 3 = ☐   4 – 0 = ☐

5 – 2 = ☐   5 – 4 = ☐   2 – 1 = ☐

5 – 0 = ☐   4 – 3 = ☐   4 – 3 = ☐

4 – 2 = ☐   3 – 2 = ☐   5 – 1 = ☐

3 – 1 = ☐   3 – 0 = ☐   5 – 2 = ☐

Write the symbol to show equal or unequal.

= or ≠

2	≠	0 + 1			5 1 + 4
1		3 + 1			3 2 + 3
3		2 + 1			4 3 + 2
5		2 + 3			1 2 + 1
4		3 + 0			2 0 + 2

On the Back Write the numbers 1–100.

Name ______________________________

1	11								
2									
10									100

Homework

Subtract the numbers. Use your fingers or draw.

4 – 1 = ☐   5 – 1 = ☐   4 – 2 = ☐

2 – 2 = ☐   4 – 0 = ☐   5 – 0 = ☐

3 – 0 = ☐   5 – 3 = ☐   4 – 3 = ☐

5 – 2 = ☐   3 – 2 = ☐   2 – 1 = ☐

3 – 1 = ☐   1 – 0 = ☐   5 – 5 = ☐

Write the symbol to show equal or unequal.

= or ≠

	10 2 + 7	8 5 + 3
	6 4 + 2	6 4 + 3
	9 4 + 4	9 2 + 5
	7 2 + 6	7 1 + 6
	8 4 + 3	5 3 + 3

 On the Back Write the numbers 1–100.

Introduction to Counting and Grouping Routines

Homework

Draw Tiny Tumblers. Write how many there are on each Math Mountain.

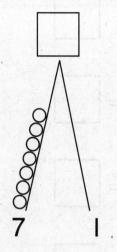   

7   1   6   2   5   3   4   4

  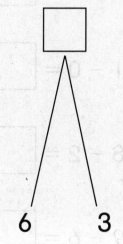 

8   1   7   2   6   3   5   4

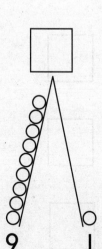   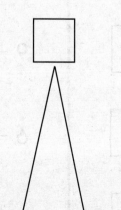 

9   1   8   2   7   3   6   4   5   5

Add Partners to Find Totals   **109**

Subtract the numbers.

$4 - 3 = \boxed{}$ $2 - 2 = \boxed{}$ $5 - 1 = \boxed{}$

$3 - 1 = \boxed{}$ $5 - 2 = \boxed{}$ $0 - 0 = \boxed{}$

$4 - 2 = \boxed{}$ $4 - 1 = \boxed{}$ $5 - 3 = \boxed{}$

$5 - 4 = \boxed{}$ $3 - 1 = \boxed{}$ $4 - 0 = \boxed{}$

$2 - 1 = \boxed{}$ $1 - 0 = \boxed{}$ $3 - 3 = \boxed{}$

$7 - 6 = \boxed{}$ $8 - 2 = \boxed{}$ $10 - 7 = \boxed{}$

$8 - 4 = \boxed{}$ $9 - 6 = \boxed{}$ $7 - 2 = \boxed{}$

$10 - 5 = \boxed{}$ $9 - 8 = \boxed{}$ $8 - 3 = \boxed{}$

$8 - 5 = \boxed{}$ $6 - 2 = \boxed{}$ $7 - 4 = \boxed{}$

$10 - 8 = \boxed{}$ $9 - 3 = \boxed{}$ $10 - 6 = \boxed{}$

Add Partners to Find Totals

Homework

Write the numbers and compare them.
Write G for **Greater** and L for **Less**.
Cross out to make the groups **equal.**

1.

| 7 | L |

| 9 | G |

2.

3.

4.

5.

Story Problems and Comparing: Totals Through 10   **111**

Subtract the numbers.

4 − 2 = ☐   3 − 1 = ☐   5 − 3 = ☐

2 − 1 = ☐   4 − 3 = ☐   2 − 2 = ☐

5 − 1 = ☐   3 − 2 = ☐   5 − 0 = ☐

4 − 1 = ☐   3 − 0 = ☐   5 − 3 = ☐

5 − 4 = ☐   5 − 2 = ☐   4 − 4 = ☐

8 − 5 = ☐   9 − 2 = ☐   10 − 4 = ☐

7 − 4 = ☐   7 − 6 = ☐   9 − 6 = ☐

10 − 3 = ☐   8 − 2 = ☐   8 − 4 = ☐

7 − 1 = ☐   6 − 4 = ☐   7 − 5 = ☐

10 − 7 = ☐   9 − 3 = ☐   10 − 6 = ☐

Story Problems and Comparing: Totals Through 10

Homework

Subtract the numbers.

$3 - 2 = \boxed{}$ $5 - 2 = \boxed{}$ $8 - 4 = \boxed{}$

$2 - 1 = \boxed{}$ $3 - 1 = \boxed{}$ $6 - 2 = \boxed{}$

$5 - 1 = \boxed{}$ $8 - 1 = \boxed{}$ $4 - 0 = \boxed{}$

$4 - 4 = \boxed{}$ $9 - 3 = \boxed{}$ $9 - 1 = \boxed{}$

$10 - 1 = \boxed{}$ $5 - 3 = \boxed{}$ $4 - 2 = \boxed{}$

$6 - 4 = \boxed{}$ $7 - 2 = \boxed{}$ $2 - 2 = \boxed{}$

$7 - 4 = \boxed{}$ $9 - 4 = \boxed{}$ $6 - 1 = \boxed{}$

$10 - 4 = \boxed{}$ $6 - 1 = \boxed{}$ $10 - 2 = \boxed{}$

$7 - 3 = \boxed{}$ $6 - 5 = \boxed{}$ $8 - 2 = \boxed{}$

➡ **On the Back** Draw two equal groups of triangles.

Subtract to Make Equal Groups